GERALDINE CONVENTO AND RANDY LYMAN

Start With Keywords

The First Step to Ranking Higher on Google

First published by Start With Keywords 2025

First edition

ISBN (print): 978-1-933181-02-8
ISBN (digital): 978-1-933181-03-5

Cover art by Tressa Whitman

This book was professionally typeset on Reedsy.
Find out more at reedsy.com

Contents

Introduction: Who, What, Why, How

Who Are You?

This book is not for marketers; it's for busy business owners who do their own marketing.

You may want to do it yourself—or you may not want to but need to, for any number of reasons. The reason doesn't matter for our purposes here.

What matters is: You're doing it yourself, or your in-house team is. You're not hiring someone outside, like a professional marketer or agency.

This book will teach you.

What Is This Book About?

It's about keywords. But what exactly is a keyword?

A **keyword** is simply the exact word or phrase that someone types into a search bar when they're searching for information on a web browser or mobile device.

Exact word or phrase. There's more to it than that, of course. But when we talk about keywords and keyword research, we're talking about the specific words being typed into a search bar.

What you'll get (like you've gotten countless times before) is a **search engine results page (SERP)** full of links to blog posts, videos, social media posts, and other types of informational content that contain the keyword you searched for. The art (and science and skill) of getting your own keyword-based content to appear higher on other people's SERPs (and thus increase its likelihood of getting clicked) is called **search engine optimization (SEO)**. We'll explain all these terms as the book progresses.

Eventually, you'll need to create that optimized, keyword-based content, which is the endgame of keyword research. But let's not get ahead of ourselves; we'll cover content in Part Three.

Keyword research sounds scarier than it is. The word "research" conjures images of tedious academic work, poring over arcane data until your eyes glaze over.

It doesn't have to be that. Keyword research simply means figuring out which exact words and phrases people are actually typing into search bars.

In business terms, it means figuring out what questions people are asking when they're looking for information, products, or services like yours.

There are many ways to do this; that's what you'll learn in this book.

Yes, you can get really deep into the weeds or lost in the forest or wherever you like to traipse. And if your business operates at scale, with hundreds of products, or publishes a dozen blog posts a day, then you almost certainly need a professional to get into the weeds for you.

Our purpose with this book is to really simplify it so that a busy business owner like yourself can get it done with minimal time and effort—so you can spend your time and energy running your business, which is what you'd rather be doing anyway.

Why You Should Care

Why should you care about keywords and keyword research?

Even if you're paying someone to do your marketing, as a responsible business owner, you need to know how people are talking about your business. Specifically, how they're asking about it, how they're looking for it—so they find your business instead of your competitor's.

Here's the kicker. In our experience as agency owners, failure to do adequate keyword research is a major reason why search engine optimization (SEO) campaigns fail, and why so many business owners end up unhappy with their marketing agencies.

For now, just understand that SEO is about questions and answers—and who can answer questions about your business better than you?

Even if you're hiring an agency, understanding keywords will help you understand what they're doing even if you don't know *how* they do it. After all, that's what you're paying them for. Understanding keywords will help you get your money's worth.

How We'll Do It (What's In This Book)

Now that you know what we're talking about and why, here's how we'll proceed:

First, this book is written by two people and we've made no attempt to try to sound like one person. We've each written particular chapters, and it'll mostly be clear who wrote which, but it doesn't matter. Geraldine and I (see, you already know who's writing this) have worked together for several years on a variety of SEO and writing projects, so you'll get the benefit of our combined experience and knowledge.

Part One

In **Part One** we lay out the basic elements of keyword research: words and people. Keyword research is all about people using words to ask and answer questions.

First, I will talk about the words. We'll put keywords in context, and cover some types and definitions. Mainly, though, we'll see how keywords relate to the all-important concept of **user intent**.

Next, Geraldine talks about the people: your audience. Or audiences; you may have more than one. They may find you in different ways and places, and be interested in reading about different things. Their interests may be common or unique, local or national or niche.

In short, by the end of Part One, you'll have a basic but thorough understanding of who and what we're dealing with, as well as a useful way to understand them.

Part Two

In **Part Two** we'll put on our gloves and go through the actual process of keyword research.

You'll learn how to create a **seed list** and then expand it using both free and paid tools. Then you'll narrow it down, using metrics like search volume, relevance, and difficulty, as well as trends and seasonality.

By the end, you'll have the basics of a keyword strategy, and we didn't even need to call it that.

Part Three

In **Part Three**, we'll cover several standalone topics, such as social media, local keywords, artificial intelligence (AI), and best (and worst) practices in SEO.

This is not a book about writing or creating content; that's a book in itself, and others cover it thoroughly. But you will learn how to cluster keywords to create what Google calls **helpful content**—the gold standard in content marketing.

We'll also talk about **paid search** (i.e. ads that you pay for), but paid search, like content creation, is a whole realm unto itself, so we won't do more than introduce it to you here. Organic search is the realm of search engine *optimization* (SEO) and of this book. Paid search is the realm of search engine *marketing* (SEM).

Finally, what would a book like this be without a section full of **Resources**? We'll cover our favorite tools and provide information on courses, downloads, and more to help you learn and improve.

I

Part One

An introduction to keywords and how they relate to people and the questions they ask.

1

Understanding Keywords

When we talk about keywords, what are we really talking about? In short, we're talking about the questions people ask.

Keyword Jeopardy

The old game show Jeopardy is a good way to understand the process of keyword research. A contestant is given an answer and must come up with the question that gives that answer.

Keyword research is the process of figuring out the questions people are asking on their browsers, smartphones, and other devices. You create blog posts, videos, and other content to answer the questions, using the same words: keywords.

Putting Keywords into Context

Keywords are the link between the questions a searcher asks and the answers provided by content on the internet. Keywords tell search engines like Google what a blog post, video, or other content is about. If you search for the keyword *apple*, Google will look for content containing the word *apple*.

In ancient times (i.e. 2010), the only way Google knew how to do this was to count the number of times a keyword appeared in a blog post: If it contained the word *apple* ten times, it was probably about apples.

So that's what you did—use the word *apple* a lot when writing a blog post. Google used a measurement called **keyword density**—the number of times per hundred words the keyword *apple* appeared—reasoning that a 500-word post mentioning apples 20 times was more likely to be about apples than one mentioning apples twice, where apples could be incidental.

You can see the flaws in this reasoning; quantity does not equal quality. What you wanted was good information *about* apples.

Which could be hard to find. You've guessed (correctly) that **keyword stuffing** was a highly profitable way for unscrupulous folk to game the system by including little to no content except the keywords themselves, just to rank higher (a so-called **"black hat"** tactic). So you clicked through to a lot of empty, worthless content in those days. Google started fighting this problem in 2012 with the so-called Penguin algorithm update. (No need to remember algorithm updates unless you're a geek like the authors.)

All you need to know is that since then, Google has not merely defeated keyword stuffing but learned to read, understand, and write natural, spoken language at least as well as most humans. Keyword density can still be a useful measurement, but it's not how people talk in real life.

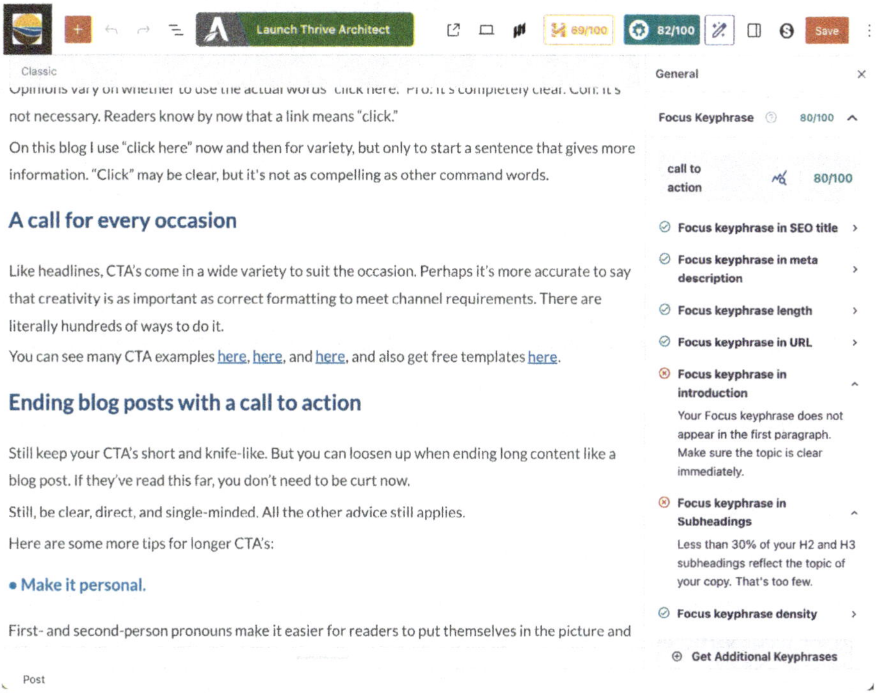

SEO plugins like Yoast, RankMath, and All in One SEO (pictured) all have a "traffic light" system of red, orange, and green marks (right side of image) to aid you in writing keyword-rich, highly optimized blog posts and other content.

How Keywords Have Changed

Today, keyword density is insignificant compared to plurals, synonyms, variations, and context—the way people talk and write naturally. The keyword density of old reads now like keyword-stuffed or AI-generated writing, and Google will devalue it appropriately.

When people search, they search semantically, which simply means: conversationally, like a question they'd ask another person. You wouldn't ask a friend, "Best pie apple?" You'd ask, "What's the best type of apple to use for baking an apple pie?"

That's what you'd ask Google, Alexa, Siri, or any other device you can ask a question. Voice search with natural language is the present and the future.

Unlike keyword density, synonyms and variations sound natural and human, and Google has always said to write for human readers first, not search engines.

But natural language still contains keywords.

User Intent: An Introduction

Keywords are at heart questions, and when people want to buy something, they ask different questions at each stage of their customer journey—a progression called **user intent** in Google-speak.

Google is all about user intent. They completely redesigned Google Analytics to track user intent, and although I have yet to meet anyone who likes the new GA4 (different story), it may ultimately prove more useful because it tracks the customer journey better.

User intent is critical to consider in keyword research. As a customer gets closer to buying, they're trying to learn different things, so they ask different questions—which means they're using different keywords.

In short: User intent mirrors the customer journey.

User Intent: An Example

I recently searched on Google and the first item on the search engine results page (SERP) was a blog post entitled: *Why You Wake Up At 4am Or the Same Time Every Night?*

My question was: "Why do I keep waking up at 4 am?" Here are some other posts that appeared:

- Reduce Stress And Stop Waking Up At 4 AM
- Here's why you keep waking up at 4am
- 4 Reasons Why You're Waking Up at 4 a.m.
- Why Do I Keep Waking Up at 4 AM? We Asked a Sleep Doctor

Here's what some other people asked:

- How can I get back to sleep at 4am?
- Why do I wake up at 4 every morning?
- Why am I waking up at 4am and can't go back to sleep?
- Why do I wake up between 3 and 4 am?
- Is my liver waking me up at 4am?

The other searchers and I were all at the start of our customer journey, looking for basic answers. We weren't ready to buy anything yet, just looking for information, some answers so we could get some sleep! All we knew was the problem. Google would say we had *informational intent.*

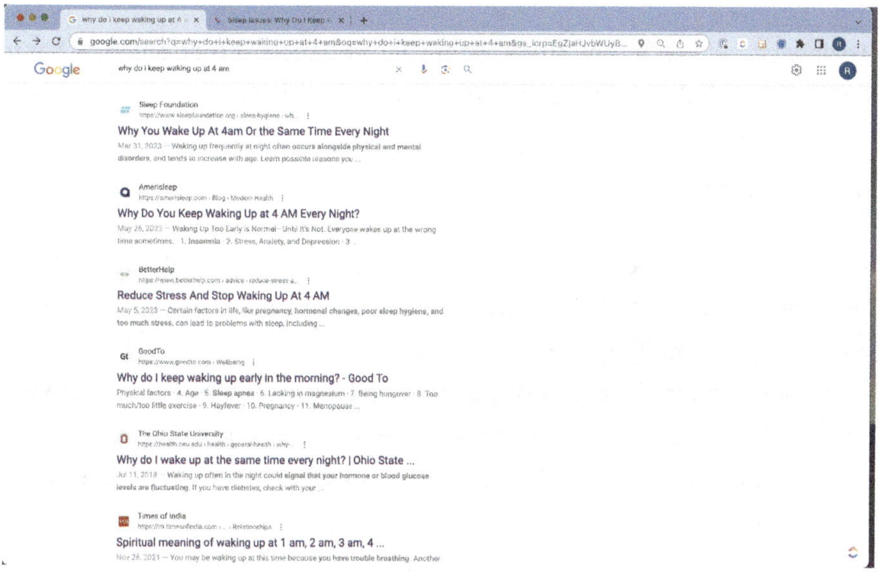

Search engine results page (SERP) for query "Why do I keep waking up at 4 am"

The point is, the blog authors all knew exactly what question—the exact words—sleepless people are asking themselves at four in the morning, just looking for some basic answers.

They knew because they did their keyword research, then created content—with our keywords in the title—that answered our questions at that stage of our customer journey.

User Intent: The Whole Journey

Let's finish the journey. Google recognizes five user intents:

1. Informational
2. Commercial
3. Transactional
4. Navigational
5. Local

To understand user intent, think about your own process for deciding what to buy. You probably go through the following stages in order:

Informational

It all starts with a need. You know the problem, but not the solution, and you may not fully understand the problem either. In short, you lack information—about the problem, the solution, all of it.

So your first task is to inform yourself, and at this stage of your customer journey, you'll be asking a wide range of general questions.

Let's go back to my question. The results we got before would be a good place to start:

- Reduce Stress And Stop Waking Up At 4 AM
- Here's why you keep waking up at 4am
- 4 Reasons Why You're Waking Up at 4 a.m.
- Why Do I Keep Waking Up at 4 AM? We Asked a Sleep Doctor

You're looking for information about the problem: what causes it, what are possible treatments—and probably learning about sleep itself: why the body needs it, what happens when you don't get enough. What kind of sleeper are you? And so on.

With *informational intent*, you're just trying to get the lay of the land. You're collecting data without much concern about where it's leading or what to do with it.

Eventually you'll reach a saturation point. You'll know enough, and feel ready to move to the next step.

Commercial

This is sometimes called *comparison intent* because the customer is basically comparison shopping: intending to buy but not ready just yet.

With informational intent you're researching the problem; with commercial intent you're researching solutions, considering options.

From a keyword research standpoint, you're asking different questions to get a different kind of answers. Your keyword-questions now include words like *best*, *review*, and *compare*.

Transactional

When you're ready to buy, you have *transactional intent*: Where can I buy the sleep mask I want?

Again, a new set of questions and keywords, e.g. *buy, sale,* and *discount*.

Navigational

Navigational intent means you know where you want to go, and you type in the business name or URL and go there. Searching for a business by name—or any search with the business name in it—is called a **branded search**.

This is what you do if you're a current client or customer, or you've decided where to buy and now just need to get to the proper website. You type Amazon into the search bar; no need to search for the *most comprehensive online shopping site* each time. Navigational intent: you navigate straight to it.

Local

Local intent is a little different, as it's not necessarily a sequential stage of the customer journey. But it is if you want to shop locally.

Nearly half (46%) of all Google searches are people looking for things in their local area. When Google detects a local intent, it employs an entirely different algorithm to determine rankings and visibility. If you're hungry for pizza; you just want the place that's closest or delivers fastest without going through the other stages.

Local intent indicates a high readiness to buy, which puts it at the end of the customer journey (**"bottom of funnel"**), and it's often indicated by keyword modifiers like "near me" and "open now."

The Long and Short of Keywords

User intent is closely related to the length of the keywords you use. Keywords come in two basic lengths:

- Short ("short-tail" "fat-head" "head term") — 1-2 words
- Long ("long-tail") — 3-5 or more words

There can be great keyword opportunities in the middle of the range (3-4 words) and these are sometimes called "medium-tail" keywords, but short-tail and long-tail are enough to give you the picture.

Let's start with a simple example, the keyword: Philadelphia

With one word, this is a head term, or short-tail keyword, a very broad category. If you search for it, you'll get top-10 results for:

- The city in Pennsylvania
- The movie starring Tom Hanks
- A brand of cream cheese

In other words, Google has no idea what you're looking for. So let's add some words:

- Sights to see in Philadelphia
- What are the cheapest flights to Philadelphia?
- Who starred in the movie Philadelphia?
- Is Philadelphia on Netflix?
- How is Philadelphia cream cheese made?

Now we're getting somewhere! These new *long-tail* keywords are going to get us some useful answers. In this instance, long-tail keywords help distinguish among broad topics (head terms) where the user's intent is unclear.

Mostly, though, lengthening a keyword like this is how you drill down into a topic or the customer journey (they're related).

- Short-tail (head term): blog
- Medium-tail (stupid term, but whatever): what is blog hosting
- Long-tail: what's the best blog hosting service

Understanding the difference between short-tail and long-tail keywords is critical for creating a keyword strategy. The riches are in the niches, as they say, and that's true of keywords as well as businesses. Long-tail keywords are niches.

They're also bottom-of-funnel. A local tour operator will never rank for a vague, general head term like Philadelphia—but it may blow its real-life competitors out of the water with "best historical walking tours in Philadelphia." And if people are searching for that, they're probably ready to buy.

Wrapping Up

In this chapter we've looked at keywords from two different perspectives: length and intent. They are related.

As your customers spiral inexorably down through their customer journey funnel, the questions they ask are not merely different at each stage, they're more specific and often longer. Long-tail keywords are more specific, closer to the bottom of the funnel.

Short-tail, head term queries like *sleep problems* become lengthy and specific queries like *what's the safest over-the-counter sleeping pill*?

The essence of creating business content is providing helpful answers to the different questions your potential customers have at each stage of their customer journey.

How do you know which questions? Get to know your audiences—in the next chapter.

2

Understanding Audiences

Defining your target audiences is a fundamental and strategic step in creating a successful SEO strategy. Many companies have three or more target audiences depending on their business model and number of products and services.

Knowing your target audience helps in drawing conclusions, such as their interests, motivations and behaviors. By pinpointing your target audience, you can tailor your SEO strategy to align with their priorities and needs. You can make more informed decisions on what search terms and phrases they may use as well as what types of content they would care about.

Here are some examples of target audiences:

- **Leads:** These are potential customers, usually grouped by various characteristics based on their needs and which of your offerings are most applicable to them.
- **Current Customer Base:** People or companies that are currently paying for your products and services.
- **Segments Within Your Customer Base:** New and returning customers, high-value clients, or niche segments.
- **Strategic Partners and Investors:** Individuals or companies that can support your business growth through collaborative offerings or funding.
- **Referral Partners:** Influencers, affiliates, or industry experts who drive referrals.

- **Press and Media:** Journalists and bloggers who can help amplify your brand's voice.
- **Teammates and Internal Stakeholders:** Employees and collaborators who shape the company's culture and operations.

By aligning your website content, keywords, and online promotions with the interests of your target audience, you can improve your website's rankings, attract qualified leads, and achieve greater success in reaching the people you need to reach.

Finding Your Target Audience(s)

There are many details about a target audience (location, income, demographics, activities, etc.) that could inform your overall marketing strategy. Together, they create a customer profile or **customer avatar**.

In the Resources section, we'll share tools that can help you dig into understanding your target audience. In the meantime, here's an activity to help you define your target audience at a high level.

To figure out who your target audiences are, ask yourself:

- What does my business need most right now—more sales, better visibility, or perhaps new partnerships?
- Who are the people that I want to reach, and what actions do I hope they will take?
- What basic details describe your ideal customers or partners, such as their age, where they live, or what they do?
- How do you think these people search for information online, and what questions might they ask?
- What are the main challenges or needs of each group, and how can you help solve them with your content?

When you understand who you're speaking to, you can use language and examples that match their interests and needs, which means that search engines are more likely to show your site to those who are most likely to benefit from it. In simple terms, aligning your SEO efforts with a clear picture of your audience makes your website more relevant and appealing both to search engines and to the people searching for what you offer.

Defining your target audience can be an in-depth task and not always intuitive to do-it-yourself. If you would like support, we highly recommend working with a marketing or brand development consultant to define all of your target audiences and determine how to best share your message with them.

II

Part Two

Step by Step: Conducting keyword research and creating a keyword list.

3

Start Your Seed List

You have to start somewhere, and the place to start keyword research is with what you already know about your business. You'll use that to create a **keyword seed list**.

In the quest of creating an effective keyword strategy, I (Geraldine) will share three distinct methods for starting a seed list.

I encourage you not only to read through the strategies, but seriously try each method yourself. Finding the right keywords is experiential. It requires trial and error because sometimes what we think people search for, actually is not it.

By experimenting with these methods, you can customize them to suit your particular needs and uncover the winning combination of words that propels your online presence to new heights.

Now it's time to play Keyword Jeopardy and discover the keywords that will make your business stand out!

Step 1: Brainstorming

Before using keyword research tools, however, you need to understand the thought process and intent behind keyword research. We covered intent in Chapter 1; now we'll cover the thought process.

The first step is to decide which approach or method to use. Experiment with one or more of the three below to create your seed list. We'll cover the remaining steps in the next chapter.

Eventually you'll need to move your work to a spreadsheet, but you can start simply with a pen and paper. For now you're just taking notes.

Download our free Keyword Research Workbook, which includes the tabs and columns you'll need to brainstorm, organize, sort, filter, expand, refine, and analyze your keyword list in this and the next two chapters. In this chapter, we're using the first two tabs.

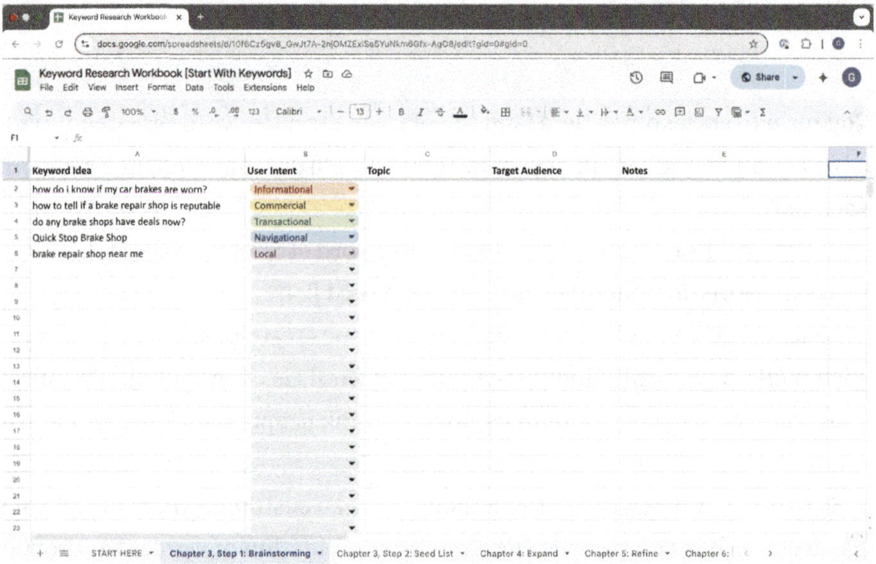

Screenshot of our free Start With Keywords workbook. Use it to create your own keyword list as you go through the steps in Part Two.

Method A: The 5 W's and How?

Start by brainstorming words that relate to your business, customers, and products or services. I recommend writing your answers shorthand and in bullet points. This will make it much easier for the keyword research tools you'll use later.

Who. These keywords represent your target audience groups that you came up with in the previous chapter. This is why you need to understand their demographics in detail: age, gender, income, etc. These are all starting points for further keyword research.

What. These keywords and phrases describe what you do. What are your products and/or services?

When. These keywords represent the time or time frame at which your audiences contact you. Immediately? While they're on vacation? Before they get married? In local search, Google now up-ranks businesses during their hours of operation. (But don't try to game the rankings by listing your hours as 24/7 if you're only open 10am-6pm; that can get you suspended.)

Where. These are location-based terms: city, county, state, region, country, or something else. Write down your business locations, but also where your customers live, work, or would make use of your products or services.

Why. These are the keywords that people type into Google that are relevant to a site like yours. What pain points are you solving for them?

How. These keywords and phrases describe the benefits of your products and services or your process of working with clients.

You can use this method alone or with a business partner who knows the business as well as you do. Remember, you're just brainstorming to start.

Method B: Question and Answer

With Method A, you start with what you already know. With Method B, you start by asking questions.

Question-and-answer is a powerful approach to crafting a comprehensive keyword strategy that resonates with your target audience. By delving into the queries and concerns of potential users, you can identify the specific phrases and terms they are likely to use when searching.

The first method you could do alone; Method B requires someone else. A good place to start is with your current clients, who will likely be happy to help—and have the distinct advantage that they already know your business.

That makes them knowledgeable enough to ask follow-up questions and engage in deeper discussion.

Networking and referral partners are another good group to tap, for largely the same reasons. Supportive friends and family make up a third circle you can explore.

Here are the kinds of things you're trying to find out:

1. What are common questions that people ask about your business?
2. What are the needs that people have in relation to your type of business?
3. If someone were looking for your products or services, what words would they type into the search bar?

The last one is the basic, core question you're trying to answer. But make your research more like a discussion than an interrogation. Remember, these people want to help you succeed.

Method C: Use Existing Data

Unlike the other two methods, this one tells you what people have actually asked, not what you think they will ask.

If your website has been around long enough and has Google Search Console and Google Analytics 4 installed, you can discover the keywords that are *already* getting your website into search results, eliminating guesswork.

Google Search Console provides a close look at how the site performs in Google's search results. Most usefully, it shows which keyword searches returned the website in search results, and which words users clicked through to the website. You'll get firsthand knowledge of exactly what they type into the search engine (even misspelled and grammatically incorrect words), and thus gain insights you can use to improve your content.

Google Analytics 4 offers detailed analytics on user behavior, traffic sources, and keyword performance over time. By examining how readers or shoppers move through your site, which sources drive the most traffic, and which actions people take (or not) on particular pages, you can gain valuable insights into what's important to your audience. Google Analytics is structured very much along user intent, which we've been discussing.

In short, Search Console shows how people are finding your website in search results, while Analytics shows what they do when they get there.

Keyword discovery is a dynamic process, requiring assumptions, hands-on research, and trial and error. By experimenting with these techniques and immersing yourself in the intricacies of your audience's search behaviors, you'll be able to customize and refine your approach. Remember, what works for one business might not work for another, making it crucial to tailor your keyword strategy to suit your unique needs.

Step 2: Create a Seed List

Creating the seed list just means organizing the notes you took using the methods in Step 1, categorizing what you and others already know about your business. It doesn't need to be long, merely to cover all the bases in a useful way. From there you can begin to discover.

Based on our experience, the co-authors find it useful to use something like the categories below to organize our seed lists. They provide a customer-centric range of opportunities for expanding the list, which we'll cover in the next chapter.

Sort the words and phrases from your Step 1 research into the following areas:

- Brand

Words with your name, your brand name, and any other names that might be associated with the business.

- Location

Write down all the places you're located and where you serve. Local businesses:

Include cities, counties, or communities. National businesses may still target states or regions, so write those down.

- Customers

Include words and phrases describing the people you serve. Be specific. You should have this from your customer profiles and demographic research.

- Pain points and concerns

What problems are your customers facing and trying to solve?

- Solutions: Products and services

List your products or services by type or category, and also by name. Branded product or service names could be under item 1 if that makes more sense for your marketing **key performance indicators (KPI)**—the numbers such as sales, new customers, or margins that are important for measuring your business success. These may differ depending on your business type or personal preference.

- Unique Selling Proposition/Niche/Process

What is unique about your business? Your process? What do you specialize in? Break this up into three categories if necessary. Identify what sets your business apart. That could be exactly what someone is searching for.

If you haven't downloaded our free workbook yet, now would be a good time. Start in Tab 1 (Brainstorming), then fill in the columns on Tab 2 for each category above. The tabs are all labeled to match the chapters of this book. You'll need spreadsheets in Chapter 5 anyway, so you might as well take a minute now to set yourself up.

Steps 1 and 2 take some work but don't need to be complicated. In the next chapter, we'll expand your seed list using a variety of approaches, and then we'll refine the expanded list after that.

4

Expand Your Keyword List

You create and build a keyword list in stages. In this chapter and the next, we'll expand and refine the seed list you created in the last chapter.

In simplest terms, you start with what's easy and work toward the more difficult. We'll cover the tools in this part of the book, but you don't need them to get started. In fact, too early and they'll probably get in the way; but by the end, they'll be really useful.

It's more important to keep in mind that your list will grow like a plant or garden: There are times to grow it and times to trim it back.

If you haven't done so already, this is the time to transfer your seed list to a spreadsheet like our free Keyword Research Workbook mentioned in the last chapter.

Step 1: Consider User Intent

As we discussed in Chapter 2 (on user intent), keywords get longer as people get closer to buying or know more about what they want.

You expand your seed list toward those longer keywords using intensifiers, modifiers, and other variables. Our workbook includes a tab full of these for reference.

When targeting readers with informational intent, your keywords should explain the what, why, and how of things. Use words like *guide*, *tutorial*, and *tips* in your long-tail keywords.

Keywords with commercial intent—when users are beginning their comparison shopping—could include words like:

- review
- versus or vs.
- compare or comparison
- best
- top

Searchers ready to buy (transactional intent) will be looking for words like:

- price
- purchase
- cheap
- buy
- coupon
- order
- deal
- special
- offer
- discount

Google can largely figure out if a search has a local intent. If you search for *best restaurants*, it will assume you're hungry and want the best restaurants nearby, not all the best ones in the world, and then engage a local search algorithm that returns results near your location.

But you can help it out by optimizing for local modifiers like *near me* or *open now* (or the opposite direction: *world's best restaurants*). We'll discuss local modifiers more in Chapter 10.

Trends and seasons offer additional reasons and opportunities to update your local keyword list throughout the year: *summer sale, back to school, spring fashion*—you get the idea.

Step 2: Study the SERPs

Another way to expand your list is with a variety of free features available in your web browser.

The place to start is with the basic search engine results page (SERP). With AI Overviews and other Google experiments, what you'll find on a SERP undergoes constant change. New features come and go like quantum particles, popping in and out of existence as Google pursues ever-changing search habits and higher profit margins.

Features like "People also ask" and "People also search for" and the "Discussions and forums" section. In Chapter 3, you asked your friends and clients which exact words they would search for. These SERP features are like asking everyone. They invariably yield long-tail keywords and previously searched terms, in natural language (another factor that's important, especially in voice search like asking Alexa).

For example, I just searched for *how to publish a book* and learned that people also asked *How much does it cost to publish a book?* and *How do first time authors get published?*

"Related searches" gave me:

- how to publish a book on amazon
- how to publish a book yourself
- how to publish a book for free and get paid for it
- how to publish a book on amazon and make money

These are all long-tail keywords—and all potential topics for a blog post and/or video. In fact, I just ran that last bullet point through some headline analyzer tools (see Resources) and it scored outstanding SEO and engagement scores exactly as is!

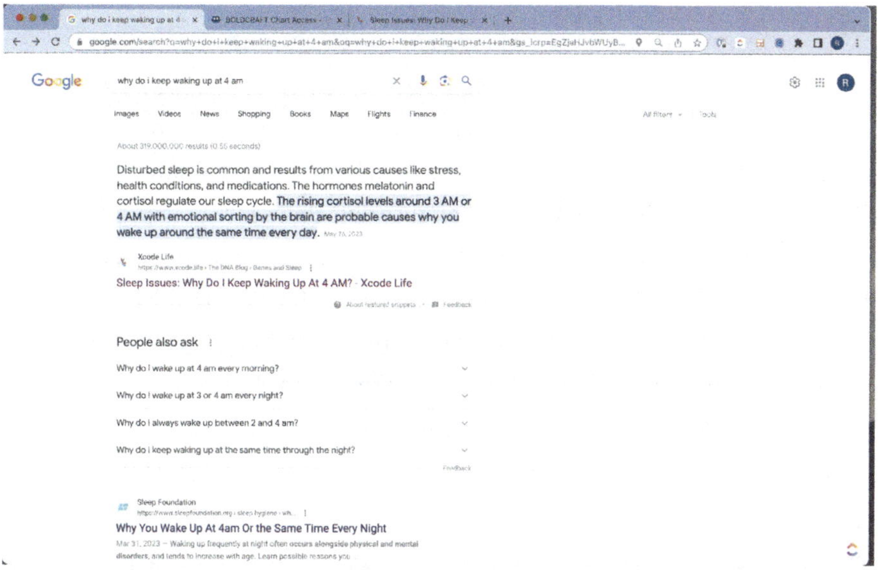

Our earlier sleep query with additional SERP features, in this case: AI Overviews and People Also Ask.

The "Discussions and forums" section returns threads from Reddit, Quora, and other forums, possibly because these reflect actual user experience, a critical feature of the "helpful content" that the search engine prizes.

I just found a Quora post—"Are there any websites that allow you to publish books for free?"—that expands my keyword substantially and could become a niche blog post title likely to have little competition in the rankings.

Step 3: Use Online Tools

There are many online tools you can use for expanding your keyword list. Using them is a crucial step in refining and validating your keyword strategy. These tools offer valuable insights and data that can help you uncover new keyword opportunities, validate assumptions, and refine your targeting.

Here are a few types of tools that you can use to gather more keyword ideas and expand what you currently have:

Keyword Research Tools

Keyword research tools such as Google Keyword Planner and Keyword.io (free), or Semrush and Ahrefs (paid), are used to explore related keywords, search volume, and competition levels. They will help fill in the columns on our seed list template (which of course you downloaded above).

These tools provide extensive databases of keywords and related terms based on historical data (previous searches conducted through search engines). They allow you to assess words based on search trends, user intent, and relevance. By inputting your initial keyword list into these tools, you can uncover additional keywords that align with your target audience's interests and search behavior.

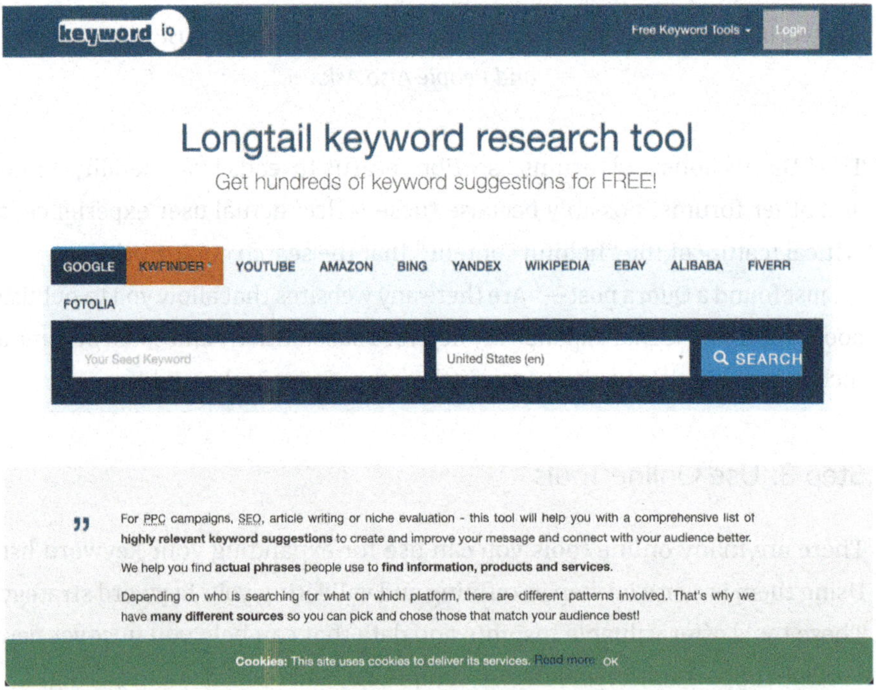

Keyword.io (that's the name and URL) is an outstanding free keyword research tool.

Auto-complete and Related Searches

Search engines and popular platforms like YouTube and Pinterest have auto-complete suggestions and related searches to uncover long-tail keywords and variations that may not have been included in your initial list.

Start typing your primary keywords into their search bars and observe the auto-complete suggestions to identify popular search queries. Also, scroll to find related searches that can provide further keyword inspiration.

Competitor Analysis

Conduct competitor analysis using tools like SpyFu or SimilarWeb (see Resources) to identify keywords that your competitors are ranking for and targeting in their own campaigns. Major SEO platforms like Semrush do this too.

Analyzing competitor keywords can reveal gaps in your own keyword strategy and highlight opportunities to target keywords that are driving traffic and conversions for your competitors. Pay attention to both organic and paid keywords to gain a comprehensive understanding of your competitive landscape.

User-generated Content and Forums

We just talked about these above. By exploring user-generated content sites and forums related to your industry or niche, you can identify common questions, topics, and keywords that resonate with your target audience. Platforms like Reddit, Quora, and industry-specific forums can be valuable sources of keyword insights, as they provide real-time feedback and discussions from actual users. Take note of recurring themes, language patterns, and specific keywords used by your target audience to inform your keyword expansion efforts.

By leveraging these online tools and resources, you can build your keyword list and make data-driven decisions based on social proof and analytics.

Continuous keyword research and analysis are essential for creating a strong keyword strategy.

5

Refine Your Keyword List

In SEO campaigns we use historical data (results of past searches) to help determine what will be effective for ranking higher in the future.

When refining your keyword list, the most important thing to remember is the meaning of **volume** and **results**. In my personal experience, I've learned that understanding this makes keyword research and list refinement much easier.

Volume (or **search volume**) represents how often a particular keyword or phrase is searched using a search engine. Most search tools give the average number of monthly searches over the past 12 months. Volume tells you how popular a keyword is, how often people are looking for it.

Results are the individual web pages that a search engine considers relevant to a keyword or phrase and displays as a link on a search engine results page (SERP).

With this understanding, I'll share the 4-step method I've developed to narrow down my keyword list. By the end of this process, I typically end up with anywhere between 25 and 300 keywords, depending on the client's needs.

The workbook for this book has sorting columns for volume and other key metrics we'll look at in this chapter.

Step 1: Identify the Range

Depending on the popularity of your industry or product, the range of results available for your keywords may vary. For example, in a highly saturated market with lots of searches and a wide variety of sites online that share information about it (e.g. movies), you'll have a wider range compared to other industries with a very niche audience (e.g. crypto surveys).

Understanding the general range of results during SEO keyword research will help you identify high-impact keywords that align with your target audience's search behavior as well as others that won't serve you as well.

Step 2: Remove the Highs

High-volume keywords are popular topics with lots of searches, which means they are highly competitive and hence more difficult to rank for. Most keyword tools have a "keyword difficulty" or "competition" column with some kind of scale they calculate—often 1 (easy/low competition) to 100 (difficult/competi tive).

It's like trying to stand out in a crowded marketplace: More people are looking, but it's tougher to get noticed because more are also vying for attention. Although high-volume keywords are relevant, they're often head terms (short-tail keywords), so you won't need many of them.

Low-volume keywords are less popular, like niche topics with fewer searches. Fewer people are searching for these terms, but that means there's also less competition, making it easier to rank higher in search results.

Balancing volume and results means finding keywords with a decent search volume (so there's interest) but not so difficult (competitive) that ranking becomes unrealistic. It's about choosing your battles wisely—focusing on keywords where your effort can make a real impact.

High-volume keywords are still important because they help search engines understand your overarching topic. Just don't hold onto all of them, they won't necessarily help you rank.

Step 3: Confirm Context

Some words have multiple meanings. For example, the word *box*. When you refer to the word *box*, do you mean a cardboard box, the online cloud storage platform Box, or something abstract like thinking outside of the box? If your brand or area of focus has multiple contexts, you'll need to review the top searches to make sure that those searches align with what you're looking to rank for. (We gave a detailed example using *Philadelphia* in Chapter 1.)

Another example of when context matters is that sometimes books are written about the exact topic where you want to rank. For example, I once worked with a client that had a focus on family business. When I looked at the most popular search results related to the keyword *family business*, there were two things that could potentially prevent me from getting that client to rank highly for that keyword: a television show called *Family Business* and a magazine called *Family Business Magazine*.

When this happens, you'll need to use other related words, as the "highs" in your keyword strategy will likely be dominated by bigger brand names.

Step 4: Conceptualize using Mediums and Lows

Now that you have your "highs"—your overall topic that you want search engines to know is related to your business—you'll need to be more specific and fill in the gaps.

Relate this to your keyword strategy like it's an essay. An essay is a focused piece of writing that explores a specific topic. It uses an overarching theme explaining something, to guide the reader through subtopics that support the main idea.

Use this same structure when coming up with your keyword strategy and putting together copy for your website.

In the end, your keyword list should have 2-5 high-volume keywords and a bunch of medium- and low-volume keywords.

Circling back to the early chapters, you can see that your high-volume keywords are also the head terms we discussed in Chapter 1.

Medium-volume keywords are subtopics. These are your medium-tail keywords.

Low-volume keywords are supplementary words that explain the fine details of the medium keywords. These (you see the pattern) are your long-tail keywords.

At the end of the day your keyword strategy needs to be thorough. It should answer every question related to your business and address every one of your audience's concerns.

This keyword strategy will be integrated into your website and can also inform your YouTube optimization, ad keywords, and social media channels, which we'll talk about in the next part of the book.

6

Where Do All These Keywords Go?

Hopefully you've been using our Keyword Research Workbook as you go through the process in the last three chapters. If so, you'll have a list of words and data that you can enter in the final tab, in the first four columns.

Now what? What do you do with all these keywords? Why do all this?

Your keyword list is the foundation of your content strategy. Now that you know what questions people are asking, you can use their words to create blog posts, videos, and other content that answers their questions—and thus drives traffic and more to your website and business. Creating what Google calls "helpful content" is a topic for another book, and there are several out there.

But there are places where it's important to use your keywords in the text and settings of a given piece of content. In this chapter we'll cover the most important of those places.

Website

Whatever else changes, content is still king when it comes to SEO, and websites are where most is still found. However well Google can read images, it's still looking for words, including the image's "alt text"—the text in the long yellow text box you see while an image is loading or the cursor passes over it—describing what's in the image.

It's important to use keyword variations and synonyms in the body text of your blog posts because that sounds like natural speech. But there are six places you should use your keywords verbatim.

In order of importance for search ranking:

1. **SEO Title.** This is usually the same as the post title (next item), but it doesn't have to be. Pretty much every website platform (WordPress, Wix, Squarespace, etc.) has a place to specify it separately from the post title. It's often under something like "SEO Settings" or "SEO and Sharing Settings." The SEO title is what Google displays as the clickable "blue link" in search results—with an exact keyword match highlighted in bold.
2. **Post/Page Title (H1 tag).** This is what appears on the web page itself, the title of your post. It's usually the same as the SEO title but there's a critical difference: The SEO title should be less than about 70 characters to not get cut off in search results; the page title can be longer. (The heading tags [H1, H2, etc.] should be applied from the formatting toolbar of your text editor so Google can recognize them as headings. Do not simply make the text larger and/or bold/italic.)
3. **Subheads (H2 tag).** These are the main divisions in your content, and Google scans them to understand what your content is about, just like you'd scan a book's table of contents to understand its organization. Google doesn't care much about keywords in H3 and H4 tags, but they can't hurt, and your readers may care. Remember the prime directive of SEO content: write for readers first, search engines second.
4. **Alt text.** Google reads this to understand your images, and it's also critical for accessibility. Alt text should include a brief description of exactly what's in the image, plus keywords like your business name and location.
5. **URL.** This is a minor factor in rankings, but a factor nonetheless. It's not necessary to include minor and connecting words, however. For example, if the title is "How to Write a Blog Post That Everyone Will Read, the URL can safely end with /how-write-blog-post-everyone-will-read.

6. **SEO Description.** This is the descriptive text that appears below the blue link on a SERP. Usually you enter it along with the SEO title in the SEO settings. The SEO description, however, is different from the "excerpt" you can specify in a blog post. The excerpt appears on the blog page; the SEO description appears on the SERP. Putting your keyword in the description will not help your rankings, but an exact keyword match (matching what the searcher typed into the search bar) will be highlighted in bold in the description (see picture), which can help your content earn the click instead of another result. If two titles both seem to answer a searcher's query, the description will likely decide which one they click.

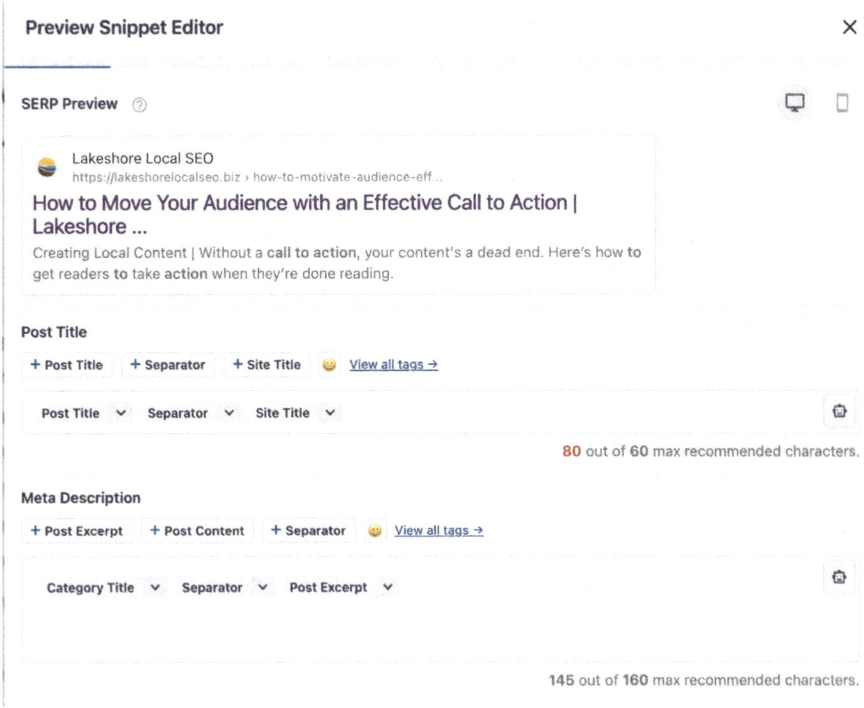

Website builders like WordPress and Squarespace have settings boxes like this to specify a title (#1 above) and description (#6) to appear in search results. Note that the focus keyword ("call to action") appears in bold in the description.

YouTube

YouTube is owned by Google, and Google searches it too. YouTube videos feature prominently on SERPs for all types of queries.

But YouTube contains its own search engine, and optimizing video is its own thing that's similar to regular SEO but with key differences.

Places to put your keyword include the video's description, which can be a transcript of the video itself; the closed captions, which are good for accessibility as well; and the tags section of the video's data box. (See **Video SEO** in Resources.)

Google Business Profile

For a local business, keywords in your Google Business Profile will not affect your organic rankings per se, but they can help in local search results (i.e. Local Pack, Local Finder, Bing Places, Apple Maps) because they show your business as relevant to a search query.

Local search employs a different algorithm that prioritizes relevance and proximity (distance from the searcher) over conventional search factors. Good places in a GBP to put keywords are your main business description and descriptions of your products and services. Encourage customers to include keywords in their reviews; naming the services they used will do.

Directory Listings (Citations)

Some online directories (e.g. CitySearch, Foursquare) may include descriptions with each listing (aka **citation**) for local businesses. These are usually pulled from the Google Business Profile. Keywords in directory listings are not a Google ranking factor; however, they're valuable in local search for establishing that your business is relevant to a local search query.

Social Media

Is SEO for social media?

We get asked that. Here's the definitive answer:

Yes and no.

Strictly speaking, search engine optimization applies specifically to search engines like Google, Bing, and a handful of others. So, no.

But in the real world, that view is too narrow and limiting. So, yes.

Because social media platforms contain their own internal search engines. There may be best practices for each that differ from those we're discussing in this book. But your keyword research can still be useful because people are searching for the same things. Refer to each platform's own guidelines for more information.

Also, Google may include results from Facebook, X, Reddit, and other interactive sites on SERPs. Social media posts do not appear in organic search results, but they can appear in carousels, video results, featured snippets, and other SERP features.

Keywords in social media are not a direct Google ranking factor, but the important word here is *direct*. Social media is an *indirect* ranking factor—or what Google calls a signal. Posting content to Facebook, Instagram, X, and other social sites can drive traffic to your website, increasing your reach and engagement, and Google considers engagement a signal that your content is useful and relevant (i.e. helpful content).

That in turn can increase the number of backlinks to your content—links from other websites back to your own—and backlinks *are* a direct ranking factor. They tell Google your content is worth linking to; the more backlinks, the more valuable your content appears. (When you link to another site, they're getting a backlink from you.)

Links from social posts are not a ranking factor or counted as backlinks, but like engagement, a large number of shares provides "social proof" that your content is of high quality, and quality is the most important ranking factor of all.

Finally, in local search, Google Business Profiles display icons to your social channels. As of this writing, Google seemed to be considering replacing its own GBP Updates feature with social media posts—meaning your social posts would appear within your GBP.

You've just gone through the process of basic keyword research for a small or medium-sized business—and now you have a keyword list.

This list can now guide you in creating content for your blog and website that answers the questions your prospective customers have when they're looking for solutions like yours—and answers them in a way that moves them through the customer journey according to their user intent, their need for new kinds of information along the way.

Our Keyword Research Workbook can be your permanent reference and workbook for reviewing and improving your keyword strategy in the future.

III

Part Three

In which we delve into various keyword topics.

7

Keywords and AI

"We are moving so quickly when it comes to AI innovations, and it's unlikely that this pace will slow down. In fact, the opposite will happen! ... AI has already changed how Google's search results work."

— "SEO in the Gemini Era," by Marie Haynes (2024)

As we're writing this (early 2025), the typical search engine results page is undergoing vast and rapid changes. Things will have changed several times again between the time we write this chapter and the time you read it. Google tests new features all the time, abruptly releasing them to limited users, then just as abruptly either canceling or spreading them.

Lately a lot of those changes are due to advances in artificial intelligence (AI)—in particular (in early 2025) the controversial AI Overviews feature that provides "result-less" answers to your search queries. And sometimes fact-less; notwithstanding, AI Overviews continues to change and improve.

The moral and ethical issues around AI are vast and ongoing, but our concerns here are purely practical and don't infringe anyone's copyright: Can you use AI for keyword research? And if so, how?

AI can indeed be a useful tool for keyword research. But the goals and thought process behind keywords (and research) are the same whether you're using AI or not.

AI & SEO Best Practices

As SEO changes, so do best practices—which is a kind way of saying that SEO is a moving target.

Nonetheless, AI is useful in many SEO applications, from brainstorming topics to writing blog and social media posts. You can identify keywords for your chosen topics, conduct research faster, generate outlines, analyze key performance indicators (KPI) like page views and content quality, find internal linking and back linking opportunities, and more. It can help you identify insights and opportunities that even an experienced pro might not see.

"Installing a new keyword research AI tool is one thing, but using it to its full potential is another," according to War Room, a digital advertising agency, which has published a brief but useful guide to thinking about AI for keyword research (see Resources).

AI has "shaped, evolved, and elevated keyword research" in several ways (the article states) by providing:

- Advanced keyword suggestions
- Predictive insights
- Content optimization
- Scalable data analysis
- Competitive analysis

AI vs. Humans

Here's the most important thing to remember in all this:

AI by itself is not enough. The human element is essential.

AI is great for generating ideas, outlines, even drafts; but to create truly original, helpful content that ranks, you need human natural intelligence (NI) and experience to finish the job. AI can enhance human effort but not replace it.

Google has always advised to write for humans first, search engines second. If you write blog posts that help people, rest assured the search engines will figure it out.

The reverse is not true. Google can fairly easily identify (and will penalize) purely AI-written content; at the same time, one of its four main criteria for excellence—for what it calls "helpful content"— is personal experience, which no AI tool can replicate.

The Art of the Prompt

The key to using AI for keyword research (or anything else, for that matter), is prompts. Prompts are just instructions: You tell an AI tool what you want. Writing good prompts is already a critical job skill for the future.

In the Resources section you'll find a list of articles full of prompts you can use for all of the SEO tasks we've listed above.

AI Tools for SEO

AI can be extremely useful for keyword research and other SEO work, and by now pretty much all SEO tools have some sort of AI built in.

Even your web browser has it, so you can get started with excellent tools like Gemini for free.

Many SEO tools have an add-on that assists with planning and creating content, where typically you pay by the word up to a limit.

So if you want to start using AI, it's unlikely you need to buy anything unless you find something you need or like more. Not all AI tools are equally good. But some paid platforms offer a lot more, like Jasper, which includes a whole suite of features for marketers.

Love AI or hate it—or both at the same time—the SEO tools of the future are already here. Might as well make the best of them.

8

Paying for Keywords

Organic vs. Paid: What's the Difference?

Business owners often come to us complaining about spending more and more money on ads but getting decreasing results.

One reason could be the lack of a strong foundation in organic search—which could boost their ad ROI. Often they confuse what paid and organic search are used for.

Throughout this book we've been talking about *organic* search results, which are results you earn by writing helpful content that answers people's questions. Thus organic results are sometimes referred to as *earned.*

But in this chapter we're introducing *paid* results—also known as **sponsored results** or **PPC (Pay Per Click)** or simply **ads.**

Whenever you hear about *paid* vs. *earned* media in almost any context, the difference is whether you pay for the exposure, or earn it because it has merit.

You'll also hear different terms. Optimizing for organic search is called SEO (Search Engine Optimization) whereas optimizing for paid is called SEM, for Search Engine Marketing. In social media too, you'll hear "optimization" used for organic and "marketing" used for paid.

Organic vs. Paid: Which is Better?

Neither is better. They're just used for different things, and often work better in tandem.

Paid search is useful when you want quick, short-term results—a seasonal sale or limited-time special offer, for example. The advantage is that Sponsored results appear above the organic blue links, so you go right to the top (or fairly close). P

The disadvantage is they only do that as long as you keep paying, although the impact gets blunted over time. So you pay more and more to get less and less.

Organic SEO is the opposite. It takes longer to show results, but they're generally more enduring. You have to earn your way to the top, but then you're more likely to stay there. At the same time, SEO is typically much less expensive to implement.

In a well-rounded strategy, they work together. In essence, SEO raises your baseline visibility, so your ad campaigns start on a higher level, which improves their ROI.

Organic vs. Paid Keywords

Paid search (SEM) requires a different kind of strategizing than organic (SEO) that's beyond the scope of this book. Other factors come into play, not least of which is your ad budget. Conversion goals are another.

Concerning keywords, the overall process of research is the same, but again, other factors enter. We've included an excellent guide to the specifics of PPC keyword research in the Resources.

In short, ads can't take the place of organic content. But used in conjunction, you can get better results for both.

How Much Does a Keyword Cost?

Paid search and digital advertising rely heavily on a metric that's not directly important in organic search: **cost per click (CPC)**. CPC is measured on search engines, social media, e-commerce platforms, websites, and mobile apps—places where you click things.

CPC shows what advertisers are willing to pay to rank for a particular keyword. When sorting a keyword list, it's great for assessing how certain keywords are used to promote products and services, as well as whether you have a chance to rank for them.

The higher the Cost Per Click, the more competition and the more advertisers are willing to pay to get their ads in front of people. However, unless you have unlimited or a vast amount of resources, these words are typically harder for a small business to compete.

Conversely, a low CPC (or a keyword that lists a CPC as zero) may suggest untapped niches or keywords with lower commercial intent.

Outside of compiling a strategy for running ads, CPC helps when deciphering competition dynamics and gauging market saturation.

Should you use Google Keyword Planner for SEO?

Google Keyword Planner is one of the first tools you'll see in lists of keyword research tools. It's a good idea to use several different tools, but Keyword Planner is a good place to start, as it's easily accessible using your existing Gmail or Google account to create a Google Ads account. You don't have to buy ads.

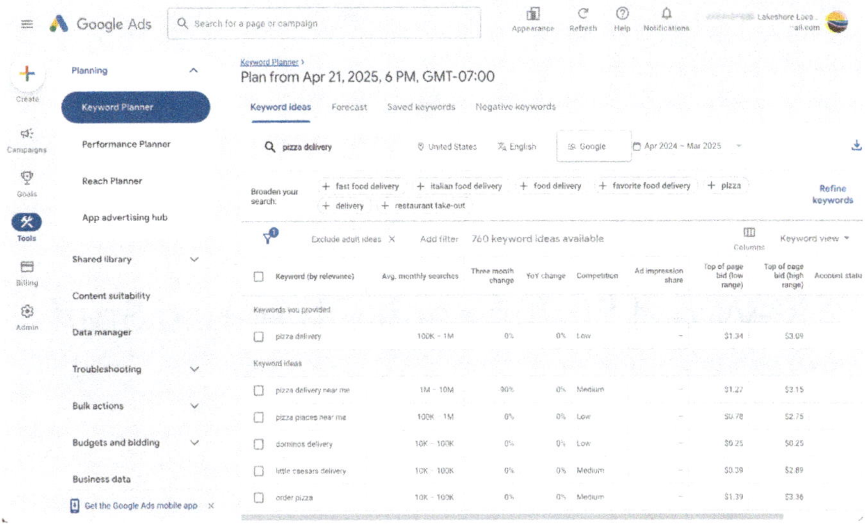

Google's free Keyword Planner tells you how much is being paid for specific keywords.

Keyword Planner provides useful information on organic search volume and other metrics, including CPC data. However, CPC shows what marketers are willing to pay for, not necessarily what people are searching for, which makes it only indirectly useful in organic search.

Still, it does show how much people are willing to pay for a keyword, and thus how desirable—and competitive—it is.

So when you're sorting your Keyword Planner results, weigh the CPC data *along with* search volume, clicks, difficulty, and the rest.

SEO is first and foremost about generating traffic to your website and building web authority. When compiling an organic keyword strategy, CPC should be more of a reference point than a deciding factor.

Finally, your paid and organic keyword research is likely to overlap a lot, but not entirely. Which means you can employ less expensive, long-term SEO to rank for key topics, and then direct your SEM spending more cost effectively to fill in holes or target more specialized topics or offerings.

9

Keyword Stuffing and Other Bad Stuff

I remember optimizing websites that I'd built in the mid-nineties. Things were a lot different back then. There were no keyword research tools like Semrush, Google Keyword Planner, Moz, etc. The internet was just the internet. Websites were found on multiple search engines, like Yahoo.

In this chapter, I'll explain keyword stuffing and so-called *black hat* SEO techniques. However, unlike other chapters where we give you step-by-step tips and tell you what to do, let me be clear now....

In this chapter, I cover what you should NOT do!

Keyword Stuffing

What is keyword stuffing?

Keyword stuffing is cramming a web page with a specific keyword as many times as possible, hoping to trick search engines into ranking it higher.

Back in the day, this had a slightly different definition. I remember the trick was to add in as many variations of the same keyword on the page as you could and color the text the same color as the background.

Either way... this is a no-no.

Keyword stuffing makes sentences sound unnatural and awkward. Search engine scripts will read, and if they notice too much of it going on throughout your website, your rankings will definitely decrease.

Black Hat SEO

What is Black hat SEO?

Black hat SEO techniques are unethical methods used to manipulate search engine rankings.

I know this sounds vague. I mean, you're just learning about SEO keyword strategy, so why would you intend to do things that are unethical, right?

Unfortunately, there are a lot of companies that practice this to achieve fast results. These are the things to be aware of when you're hiring people who implement SEO.

Here are some examples of Black Hat SEO techniques:

- **Cloaking.** Creating website pages that serve one set of content to search engine crawlers (like Googlebot) and another set of content to human users who visit the site.
- **Doorway Pages.** Creating low-quality pages with little to no valuable content optimized for specific keywords to funnel traffic to a main website.
- **Paid Links.** Buying or selling links solely for the purpose of manipulating search engine rankings, which violates search engine guidelines.
- **Private Blog Networks (PBNs).** Creating a network of interlinked websites solely for the purpose of manipulating search engine rankings through artificial link schemes.
- **Content Scraping.** Copying content from other websites without permission and republishing it to manipulate search engine rankings or generate traffic.
- **Article Spinning.** Automatically rewriting existing content to create multiple versions with minor variations, often resulting in low-quality, nonsensical content.
- **Clickbait.** Using misleading or sensationalized headlines to attract clicks, often leading to a high bounce rate and negative user experience.

- **Keyword stuffing.** Also considered a Black hat technique.

By avoiding all of the things listed in this chapter, you'll have better search engine visibility and avoid being banned from search engines.

10

Local Keywords

With local keyword research, you follow the same process of creating a seed list that you expand and refine. But as with paid search, now there are additional considerations. Local keyword research is better described as local-*ized* research.

Local Keywords: Inclusive vs. Exclusive

The difference with local SEO is that now your keywords need to reflect your business location as well as your topic.

In compiling your seed list, you should have listed the cities, towns, counties, or other geographic areas you serve. So you have what you need in order to create localized versions of your existing keywords.

Keywords that include local modifiers are called *inclusive* keywords. As opposed to *exclusive*, without them. Modifiers can include place names or phrases suggesting proximity, like *near me* or *open now*.

Moving company is an exclusive keyword; *moving company near san francisco* is an inclusive keyword.

If you're doing keyword research for a local business, include both inclusive and exclusive keywords–with and without local modifiers–in your SEO strategy and monitoring.

Does Google Business Profile need keywords?

Yes and no. Static features of your Google Business Profile are a good place to include the same keywords you're using on your website.

Keywords are not a ranking factor *per se* on a Google Business Profile the way they are on a website, they won't improve your rankings. But they can still help boost your visibility in local search results.

What they do is provide signals to the local search algorithm that your business is relevant to a local searcher's intent, i.e. whether you provide the product or service they need in their area. So use keywords where you can.

Good places to include both inclusive and exclusive keywords in your GBP are:

- **Business description.** This is the longest single body of text on the GBP, so make good use of it.
- **FAQ.** Create one by using the "Ask a question" feature to ask and answer your own questions, ones that people are always asking you.
- **Product and service descriptions.** Adding descriptions to your listed products and services is known to improve local rankings; including keywords helps.
- **Google posts.** Posting an update, offer, or event once a week keeps your profile content fresh and localized.
- **Reviews.** You can't review your own business, but you can encourage happy customers to name the product or service they used in their reviews; those count as keywords.

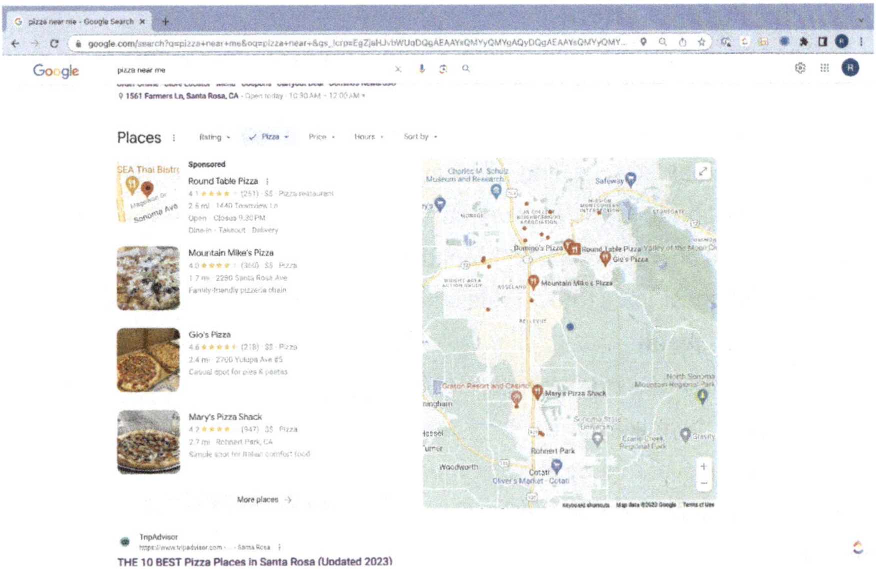

Local pack (aka map pack) for the "inclusive" keyword: pizza near me. Local modifiers like "near me" make a keyword inclusive.

GBP Keyword Stuffing

Do *not*, however, put local, service, or amplifier keywords in your GBP business name unless they're actually part of your legal business name.

If *San Francisco Plumbing Solutions* or *Bagels of New York* is actually the name of your business—that's how it's printed on your state contractor license, business registration, storefront, business cards—well then that's totally legit. But *Hilltop Plumbing – Best emergency plumbers in San Francisco, open 24/7* is a violation.

Yes, lots of people don't get caught. But they're always at risk. Stuffing keywords into your business name violates the policies of all search engines and can get your Google profile suspended without warning. Reinstating your profile can be really expensive and time-consuming.

Creating Local Content

Content pages—especially service and location landing pages—are exactly the right place to put inclusive keywords from your research, keywords that name the product, service, and/or location that is the focus of the page.

Blog posts about local features and activities—street fairs, community events, or local businesses and personalities—also help to verify your business location and boost your topical relevance, two of the most important factors in local search results.

Include local names in your posts and pages. Name cities and neighborhoods, and also local business names.

Linking to a local business from your website gives you an opportunity to ask for a link in return (backlink). Use the other business name as your link text (aka *anchor text*).

Unlike regular search, local backlinks are more important than high-authority backlinks in local search. So your content can help establish your business location and prominence—two of the three critical local search factors (along with relevance).

Last Word

Local SEO is a branch of SEO that takes into account the additional factor of location. When Google detects local intent, its local search algorithm looks for what's close above all. When you ask it for *pizza*, it assumes you're not hungry, but not for knowledge.

Local keyword research needs to tell searchers critical information like where you are, when you're open, and what you do.

11

Monitoring and Adapting

The great thing about search engine optimization is that you're able to track its effectiveness over time. It's usually best to have a significant amount of data to analyze, months or years as opposed to a few days or weeks. Depending on your goals and how active you are in updating your website content, a significant amount of data could mean that you review your analytics quarterly, every six months, or yearly.

When my (Geraldine's) team and I are working on a marketing campaign with clients, we assess search traffic, position, and the keywords that their websites are coming up for quarterly. That gives enough data and time to make adjustments or double down on certain things we've implemented for the next quarter.

Search is based on what users are searching and how valuable your website content is compared to other websites. It's important to recognize that SEO is dynamic; your website's position in search results will naturally fluctuate over time. By monitoring performance metrics, and adapting your strategy accordingly, you can maintain and even improve your search visibility as you go along.

Remember, SEO success is not just about achieving high rankings for specific keywords; it's also about delivering valuable content and exceptional user experiences that resonate with your audience—the gold standard of Google's so-called "helpful content." By continually refining your approach based on data-driven insights, you can position your website for long-term success.

How to Track Metrics

There are numerous free and paid tools available for tracking website performance through search engines. At a minimum, install Google Analytics and Google Search Console, because you can't tell much without them. They're the absolute basics, and they're free.

Google Tag Manager can be useful if you have a number of products or actions throughout a website that would be advantageous to track. However, since it's complex and requires in-depth setup, I recommend working with someone who's familiar with Google Tag Manager to discuss your tracking strategy and help you implement it.

Before moving forward with paid analytics tools, I recommend considering what your entire marketing and online presence budget is, then considering the ROI (Return On Investment) for using the tool before investing.

What Metrics You Should Track

What you track depends on your goals. For an ecommerce business, you might focus on a combination of metrics like product page traffic, shopping cart abandonment rate, top-converting keywords, and revenue attributed to organic search.

For instance, tracking which search terms lead users to your best-selling products can help you refine product descriptions and ad campaigns. This information could be pulled from whichever source you use to track website analytics as well as your ecommerce tool like Shopify, BigCommerce, etc. Different tools track different things, so keep that in mind.

A service-based business, on the other hand, might prioritize contact form submissions, phone call clicks, or bookings that originate from search traffic.

Tracking which blog posts or service pages lead to the most inquiries gives insight into what topics your audience cares about—and what converts them. If your goal is visibility, then track impressions and keyword rankings. If your goal is leads or sales, focus on traffic to high-intent pages and how often users take the next step.

Let your objectives determine your key performance indicators (KPIs), so your SEO strategy always supports your bottom line.

Tracking KPIs (Key Performance Indicators)

A KPI, or Key Performance Indicator, is a measurable value that shows how effectively you're achieving a specific goal.

In the context of SEO, KPIs help you understand whether your efforts are driving real business results—such as visibility, traffic, or conversions. For example, tracking your keyword rankings over time shows whether your content is becoming more discoverable; while monitoring bounce rates (the percentage of visitors who leave your site after viewing only one page) or time on each page helps you evaluate how engaging your site is for visitors.

By selecting KPIs that align with your business goals—like contact form submissions for a service-based business or completed checkouts for an ecommerce store—you can make data-informed decisions and adjust your strategy for maximum impact.

In either scenario, if you're not achieving results that align with your goals, you'll know that you may need to make some changes to your strategy or add more content.

12

A Few Final Key Words

Our intention for writing this book was to create the first practical, usable guide to conducting keyword research for business owners who don't really have time to do it but still need to.

Keywords are the foundation of search engine optimization. They are the words that describe your business and answer the questions your (potential) customers are asking.

We've seen SEO strategies fail because they did not get their keywords right at the beginning.

So we've discussed the foundational role that a strong keyword strategy plays in connecting your business' products and services with your clients and customers. By understanding how people search and what they search for, you can ensure that your business is visible when people are looking for it, when it matters most.

Throughout the chapters, we've broken down the process of keyword research, from the basics of identifying relevant terms to the complexities of refining and applying them in your content strategy.

As you move forward, keep in mind that SEO is an evolving field, and tracking your metrics as we just discussed is key to keeping your business visible. By observing how internet audiences are finding you—and which keywords they're using—you'll have a clear path to understanding what's working, what is not, and what may be missing from your strategy.

The tools, strategies, and best practices discussed in this book are just the beginning; hence the title, *Start with Keywords.* Search algorithms favor websites that regularly update and expand their content, seeing them as more active and engaged. By consistently producing helpful content, you will keep your audience engaged and signal to search engines that your business provides useful information and is authoritative in its field.

So start planting your seed list and see how it goes!

Resources

"Start With Keywords" – Free Keyword Research Workbook

We've created a **free Google Sheets workbook** to use alongside this book as you conduct your own keyword research. Click and you'll be prompted to make a copy that you can rename and customize as you need.

https://docs.google.com/spreadsheets/d/10f6Cz5gv8_GwJt7A-2njOMZExiSe5YuNkm6Gfx-AgO8/copy

You can find all the resources below on the Start With Keywords website at **StartWithKeywords.com/resources**. All the links worked at the time of publication. Probably they still do. No guarantees though.

Articles

User Intent

Moz: "Search Intent and the Organic Funnel"
https://moz.com/blog/search-intent-organic-funnel-whiteboard-friday

Search Engine Land: "An SEO guide to understanding user intent"
https://searchengineland.com/seo-guide-user-intent-429687

Market Muse: "What is User Intent?"
https://blog.marketmuse.com/what-is-user-intent

Keyword Research

Semrush: "Keyword Research for SEO: What It Is & How to Do It"
https://semrush.com/blog/keyword-research

WordStream: "Easy Keyword Research Guide"
https://wordstream.com/blog/ws/2017/08/09/easy-keyword-research-guide

WordStream: "PPC Keyword Research Guide"
https://wordstream.com/blog/ws/2013/11/21/ppc-keyword-research-guide

AI & Keyword Research

HubSpot: "17 Best AI SEO Tools & How to Use AI in 2025"
https://blog.hubspot.com/marketing/ai-seo

War Room: "The Guide To Using AI For Keyword Research"
https://warroominc.com/institute-library/blog/guide-to-using-ai-for-keyword-research

Surfside PPC: "6 Awesome ChatGPT Prompts For Keyword Research"
https://surfsideppc.com/chatgpt-keyword-research

AI Monks: "25+ ChatGPT Prompts for Keyword Research that Save Hours"
https://medium.com/aimonks/25-chatgpt-prompts-for-keyword-research-that-save-hours-8886cfcdb678

Narrato: "30+ ChatGPT Prompts for SEO and Keyword Research"
https://narrato.io/blog/30-chatgpt-prompts-for-seo-and-keyword-research

Morningscore: "The BEST ChatGPT prompts for keyword research"
https://morningscore.io/chatgpt-keyword-research-prompts

SEO AI: "Top 40 ChatGPT Prompts for SEO Professionals"
https://seo.ai/blog/chatgpt-prompts-for-seo

Video SEO

Google Search Central: "Video SEO best practices"
https://developers.google.com/search/docs/appearance/video

Backlinko: "Video SEO: The Definitive Guide"
https://backlinko.com/video-seo-guide

Semrush: "Video SEO: How to Get Your Videos Ranked on Google"
https://semrush.com/blog/video-seo

Social Media

Semrush: "Social Signals & SEO: Do Shares + Likes Help You Rank Higher?"
https://semrush.com/blog/social-signals-seo

SEO Tools

Keyword Research Tools (free)

Google Trends
https://trends.google.com
Google's free tool is a great place to start for seeing the big picture on trends, search volume, geographic data, and more.

Answer the Public
https://answerthepublic.com

This popular tool tells you what people are asking on Google, Bing, YouTube, Amazon, and TikTok. A free account gets you three searches a day, paid accounts more.

Keyword.io

https://keyword.io

This free tool generates keyword ideas from Google, Bing, YouTube, Amazon, and half a dozen more sites.

SEO Plugins

Plugins connect your website to Google Analytics 4 and Google Search Console so you can view their data inside your website dashboard. And it's no longer just WordPress; Squarespace and Wix have grown extensive SEO capabilities and plugin marketplaces too.

Yoast

https://yoast.com

This is the original free SEO plugin and probably the most popular too. There's a separate paid plugin add-on for local SEO.

RankMath

https://rankmath.com

This plugin has everything you could want, but it's probably best suited for website developers who know what all those settings mean. Upgrade to paid for local SEO features.

All In One SEO

https://aioseo.com

Like Yoast and RankMath, AIOSEO offers a full set of optimization tools. Upgrade to paid for local SEO. MonsterInsights (monsterinsights.com) is a Google Analytics plugin by the same company, and together you can get a complete view of your SEO and analytics inside your WordPress site.

SEOSpace

https://seospace.co

A general SEO plugin just for Squarespace users! Start for free and upgrade inexpensively.

SEO Pro Tools

These are all comprehensive platforms with an extensive set of search optimization and marketing tools that do a lot more than keyword research. When you hire an agency, they're using these or an in-house platform that does the same or similar things.

Semrush

https://semrush.com

Moz

https://moz.com

Ahrefs

https://ahrefs.com

SE Ranking

https://seranking.com

Competitor Research

SpyFu

https://spyfu.com

The pro tools can do a lot of this, but SpyFu is designed specifically for in-depth competitor keyword research in both organic and paid (PPC) search.

SimilarWeb

https://similarweb.com

Powerful and not inexpensive tools for comprehensive competitor research, including company research, shopper analysis, and more.

Local SEO

BrightLocal

https://brightlocal.com

Whitespark

https://whitespark.ca

Moz Local

https://moz.com/products/local

Local Falcon

https://localfalcon.com

Browser Extensions: Keywords & SEO

These extensions for the Chrome web browser (they work in Microsoft Edge too) let you conduct a lot of keyword and competitor research for free, and a few have inexpensive paid plans that add a lot of features. They're a good way to get started with SEO tools without the expense or learning curve of the pro tools.

Ahrefs SEO Toolbar

Lots of free SEO features, and more with a paid Ahrefs account.

https://chromewebstore.google.com/detail/ahrefs-seo-toolbar-on-pag/hgmoccdbjhknikckedaaebbpdeebhiei

Detailed SEO Extension

Built by SEO pros for daily analysis and insights on any website.

https://chromewebstore.google.com/detail/detailed-seo-extension/pfjdep

jjfjjahkjfpkcgfmfhmnakjfba

gs location changer

Be somewhere else; get search results for different locations.

https://chromewebstore.google.com/detail/detailed-seo-extension/pfjdep jjfjjahkjfpkcgfmfhmnakjfba

Keyword Surfer

Generate keyword ideas and see search volumes directly in Google search results.

https://chromewebstore.google.com/detail/keyword-surfer/bafijghppfhdp ldihckdcadbcobikaca

Keywords Everywhere

Get keyword search volume, CPC, and competitor data for more than a dozen major search sites. Free and paid plans.

https://chromewebstore.google.com/detail/keywords-everywhere-keywo/ hbapdpeemoojbophdfndmlgdhppljgmp

MozBar

Lots of free SEO features, and more with a paid Moz account.

https://chromewebstore.google.com/detail/mozbar/eakacpaijcpapndcfffdg phdiccmpknp

Open SEO Stats

Quick access to web rank and SEO stats, link stats, structured data info, and more.

https://chromewebstore.google.com/detail/open-seo-statsformerly-pa/h bdkkfheckcdppiaiabobmennhijkknn

SEO Minion

Helps with daily SEO tasks such as on-page analysis and broken link checking.

https://chromewebstore.google.com/detail/seo-minion/giihipjfimkajhlcili pnjeohabimjhi

SEOquake

Free browser plugin with key metrics, SERP analysis, audits, and other tools.

https://chromewebstore.google.com/detail/seoquake/akdgnmcogleenhbcl ghghlkkdndkjdjc

Ubersuggest

Get keyword search volume, CPC, and other insights for Google, YouTube, and Amazon.

https://chromewebstore.google.com/detail/ubersuggest-seo-and-keywo/ nmpgaoofmjlimabncmnmnopjabbflegf

Browser Extensions: Local SEO

These powerful local SEO browser plugins provide a wealth of insight and yet are inexpensive and easy to use. Both are great for competitor research as well as insight into your own business.

GMB Everywhere

gmbeverywhere.com

Operates exclusively as a browser extension that adds a wealth of Google Business Profile data to Maps and SERPs while you're logged in. Free and paid plans.

https://chromewebstore.google.com/detail/gmb-everywhere-gbp-audit/o ibcaeeplepnjfjhokfcabnaafodppik

PlePer

pleper.com

Browser extension and website with a range of tools for analyzing your business and your competitors in local search. Paid plans only.

https://chromewebstore.google.com/detail/pleper-local-seo-tools/cpdkch

kkmcpnfgbpjjgfhcbcaceeandd

Headline Analyzers

These tools help you craft perfectly optimized headlines using the best word choices for both search (keywords) and engagement (action/power words).

Headline Analyzer by CoSchedule

https://coschedule.com/headline-analyzer

CoSchedule's Headline Studio is one of the best paid tools available for optimizing headlines, but its free tool analyzes your headline ideas for blog posts, social media, videos, ads, Tik Tok, and more. Requires a free account to get your results and a limited-use version of the full Studio. Includes a browser extension and website integrations.

All In One SEO (AIOSEO) Headline Analyzer

https://aioseo.com/headline-analyzer

Good, simple, free headline analyzer from the maker of one of the leading SEO plugins. Includes word banks to help you improve your headline and try again.

MonsterInsights Headline Analyzer

https://www.monsterinsights.com/headline-analyzer

Same as the AIOSEO tool. MonsterInsights and AIOSEO are owned by the same company.

HubSpot Free AI Headline Generator

https://www.hubspot.com/campaign-assistant/ai-headline-generator

Tell HubSpot what you want and it will AI-generate an optimized headline for your blog or social post, ad, or whatever. You'll be prompted to create a free HubSpot account.

Made in the USA
Monee, IL
03 November 2026

67383032R00046